Learning with Teddy Bears

Written by Elizabeth Graham

Illustrated by Angela Munro

Claire Publications

Published by:

Claire Publications
Unit 8
Tey Brook Craft Centre
Great Tey
Colchester
Essex CO6 IJE

ISBN 1 871098 27 0

Printed in Great Britain

Introduction

When a group of children play with teddy bear counters the ideas and learning generated never cease to amaze me.
The creativity is wonderful and it is this that has prompted me to write this, my second book, about teddy bear counters.

The activities in the book are cross curricular and involve science, technology, language, with mathematics the predominant subject. They are open ended and can be adapted to suit various abilities and age ranges, although they are intended for children aged 6-8.

As you watch and work with the children on these activities, many more ideas will emerge, and much more exciting learning will take place.

Table of Contents

How many teddies do you need to cover the spots on this butterfly?

Estimate first.
Now put teddies on and check.

Work with a friend.

1. Cover your butterfly in red and green.
   Make it symmetrical.
2. Hide it.
   Tell your friend how to cover it in the same way you did.
3. Make your butterfly with yellow, blue and green teddies.
   Use the same number of each.

With another friend:

Cover the butterfly with the same number of green, blue, yellow and red teddies.
Make it symmetrical.

Cover one wing. Make the other wing the same.

1. Can you make a butterfly with half red and half blue teddies?
   How many teddies are in half of her?
2. Can you make her look different?
   Without changing the teddies?
3. Can you make her a quarter red,
   quarter green, quarter blue, and
   a quarter yellow?
   How many teddies in
   a quarter of her?

With some friends
Cover a quarter in red,
a quarter in yellow,
a quarter in green,
a quarter in blue.
Each one of you make a
different pattern.
Make a beautiful butterfly
display.

Ten teddies went to the park. They all wanted to be with a friend. Where did they play?

Change them. Don't push them with the same teddy each time.

1. Change again.
   Put three teddies on the roundabout, four on the slide, one on the swings and the rest on the climbing frame.
2. Change again.
   Put five teddies on the swings and the rest anywhere.
3. Change again.
   Put seven teddies in the park and the rest watching.

Record your work.

Change again.
Put
Ten teddies in the park
but three less on the swings than the roundabout.

Fifteen teddies anywhere you wish.
Record your work.

Twenty teddies in the park but one more on the swings than the slide and two less on the slide than the roundabout.

Teacher Notes

The previous sheets offer work on counting, fractions, symmetry, shape and area.

THE ACTIVITY ON BUTTERFLY SHEET 1
asks children to estimate, count and then work with halves and quarters.

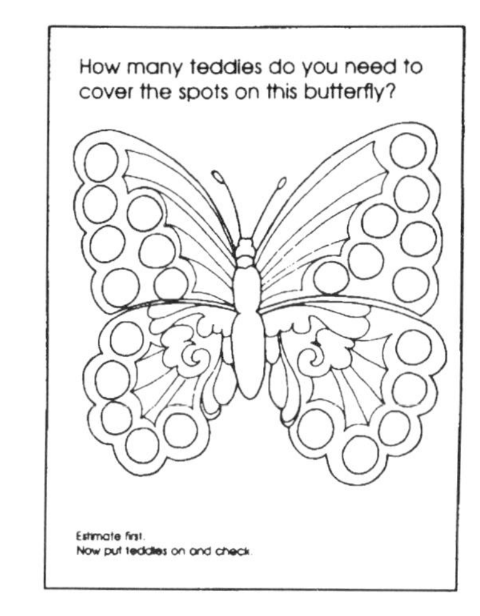

- The initial covering will be simplistic, e.g. the top half red and the bottom half blue. What is important is for children to understand that no matter how the teddies are arranged, half is still a half.
- A large display of halves coupled with discussion will help develop this concept.

FURTHER SUGGESTIONS/EXTENSIONS

- Try covering a half and leave the other half uncovered.
- Cover a quarter and leave the rest uncovered.
- Recording can be on the sheet.The only criteria is that it is a clear explanation.

THE ACTIVITY ON BUTTERFLY SHEET 2
gives experience with symmetry as well as helping to develop an understanding and use of mathematical language.

Cover one wing. Make the other wing the same.

- The point here is to imagine the wings of the butterfly closing and the patterns on the wing touching, as they might on a real butterfly. This could be developed by looking at pictures of butterflies or just as an exploration of pattern without any constraints on the arrangements.
- Giving instructions can be difficult and comments, like 'Put a red teddy there', is a favourite start. Children need plenty of practice to understand that instructions should be precise enough for them to understand without being shown.
- Work can be recorded on the sheet itself by making and colouring symmetrical patterns.

FURTHER SUGGESTIONS/EXTENSIONS

- The teacher can cover the instructions when photocopying and write their own.
- The sheets can be used for counting:

  How many? How many per wing? How many pairs? How many threes? etc.
- Give specific instructions for covering:

  e.g. Cover, but don't put the same colour next to each other.

  Cover, but make sure that each teddy touches one of it's own colour and another colour.
- Real butterflies can provide wonderful patterns to copy.
- Children can make their own creatures to cover.
- Children create their patterns and then ask friends to work out the rule they used.

THE ACTIVITY ON THE PLAYPARK SHEET

asks children to place teddies on attractions in various number groups whilst beginning to work on operations, patterns, rules and algebra.

FURTHER SUGGESTIONS/EXTENSIONS

- Teachers can use the worksheet for their own activities by blanking out the instructions and replacing them with their own.
- Increase/decrease the number of teddies to be worked with.
- Put teddies on attractions and ask children to count and make a chart.
- Give different instructions:

  Put an odd number of teddies on each attraction.
  Put even numbers on each attraction.
  Put different coloured teddies on each attraction.

Draw around your hand. Measure with teddies. Draw a graph of the information.

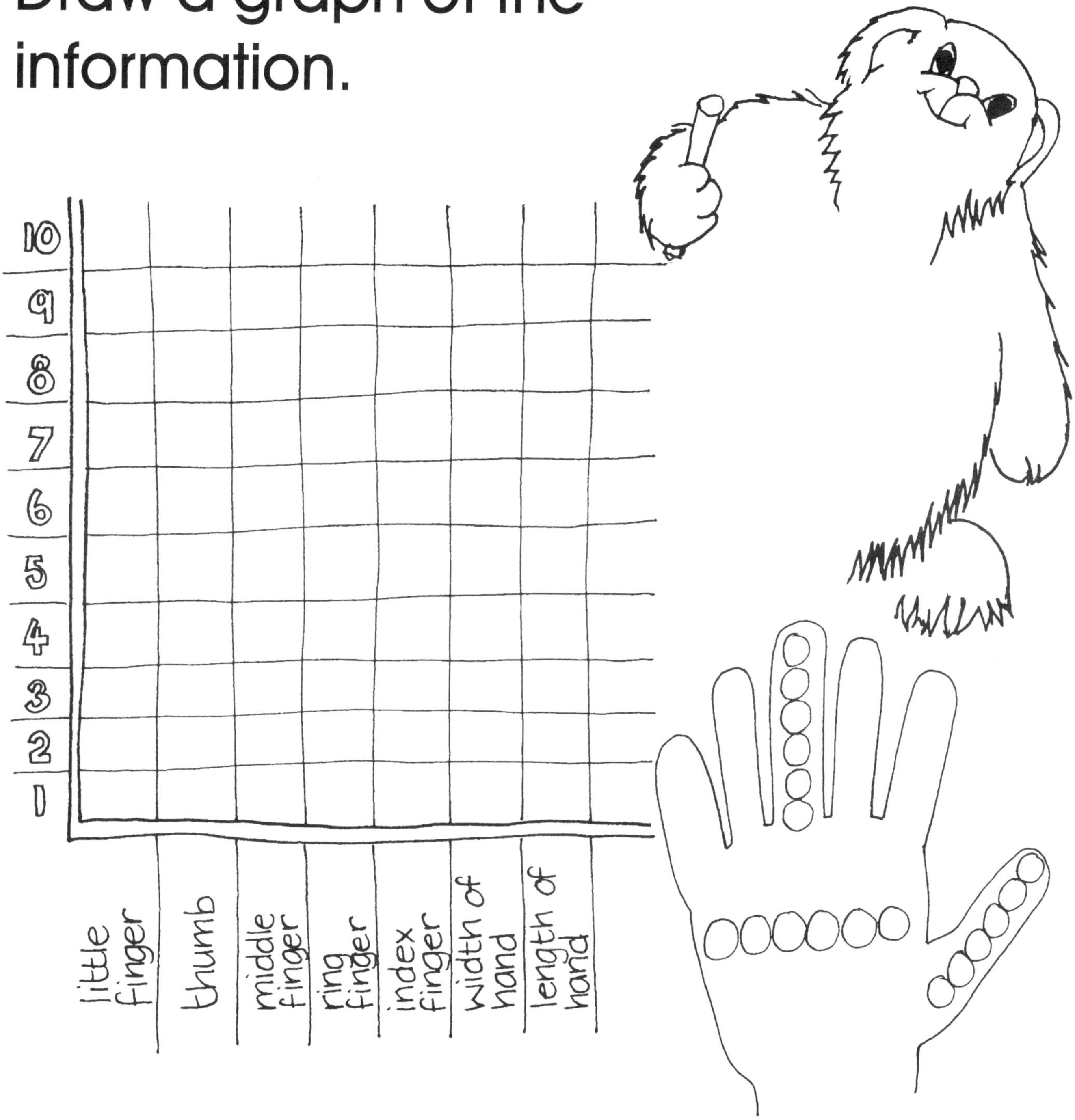

10
9
8
7
6
5
4
3
2
1

little finger
thumb
middle finger
ring finger
index finger
width of hand
length of hand

Give your graph to a friend.
Ask him or her to draw your hand using your graph.

1. Compare your outline with a friend.
2. Cover this giant hand with teddies and make a graph of it.

I wonder ...

If other friends do a hand graph can your class try to match up the hands with their graphs?

Take a piece of paper.
Roll it up to make a cylinder.

Make the cylinder fat.
Guess how many teddies will fill it.
Check.

1. Make a thin cylinder. Guess how many teddies fill it now. Check.
2. Which holds the most teddies? By how many?
3. Glue the cylinder which holds the most teddies.
4. Glue the cylinder which holds the least.

Can you:
make paper cylinders for
ten teddies
fifteen teddies
twenty teddies?

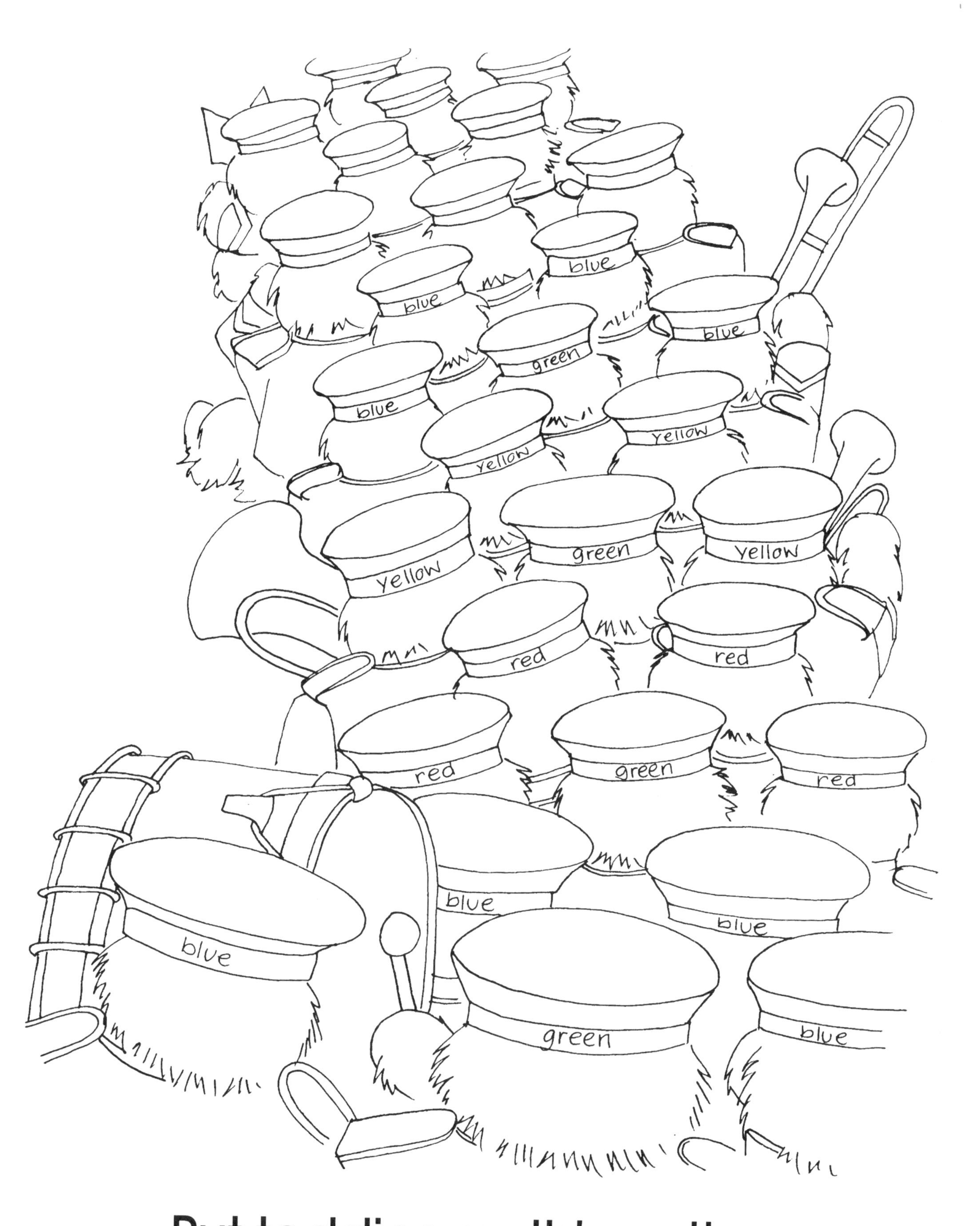

Put teddies on this pattern. Continues across your table. What will the last teddy be?

1. Guess how many of each teddy you have used. Count.
2. Start another pattern with twenty teddies.
3. Ask a friend to continue it across the table. Which teddy will you use last?

Make other teddy patterns
with twenty teddies.

Ask friends to continue them.

Find ways of recording all the patterns.

Value Teddies in Hundreds, Tens and Ones like this;

Blue and green are 1

Red is 10

Work with a friend

Yellow is 100

With a friend close your eyes and take a handful of teddies.
Using the values sort out your teddies in numbers
from smallest to largest.
What are they worth altogether?

Play this game with 5 friends. One is the banker and looks after the teddies.

Value the teddy e.g. blue = 1, red = 10.
Each player needs a hundred square.
Take it in turns to spin both spinners.
Take the number of teddies you spin.
e.g. 3 reds or 4 blues.
You can exchange
e.g. 10 blues for 1 red, if blues are ones and reds are tens.
Make numbers with your teddies and cover that number on your hundred square.
First one to get five in a row wins.

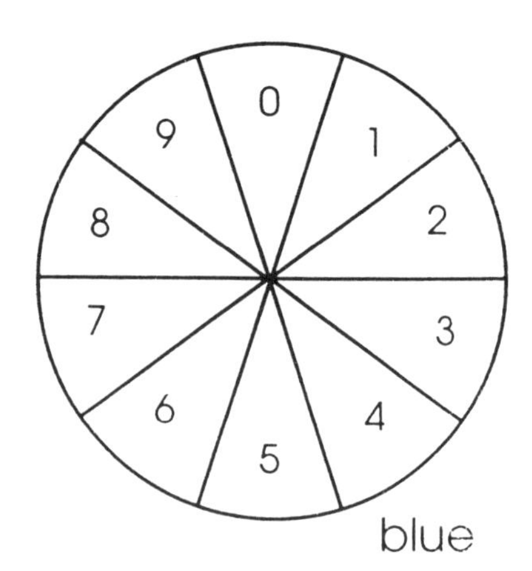

How to use the spinner:
You need:
1. a pencil
2. a paperclip
Place a paperclip on the star at the centre of the spinner put the point of the pencil on the star inside the paperclip, flick the paperclip once and see where it stops.

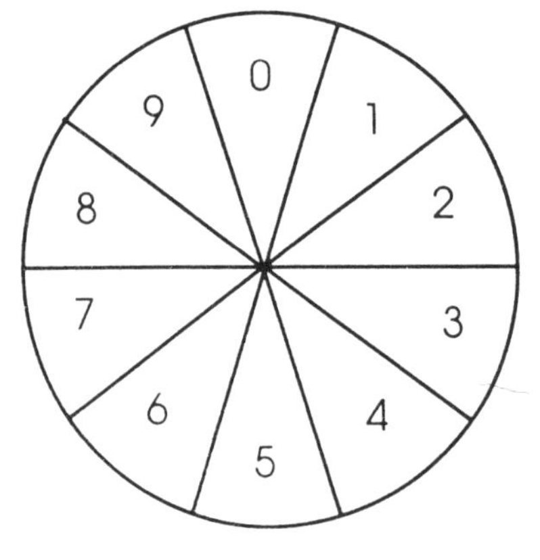

red

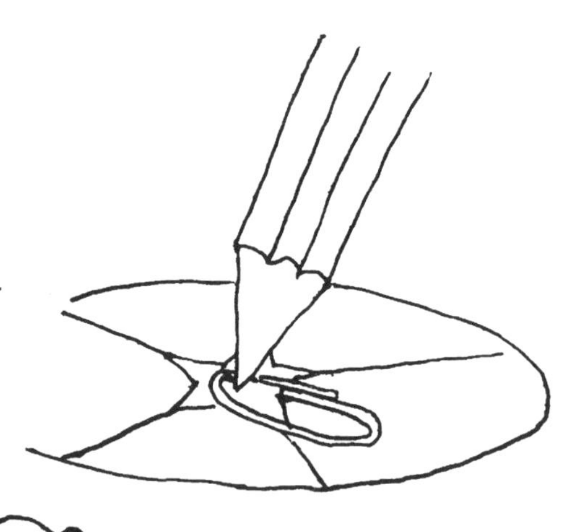

Teacher Notes

The previous sheets involve numbers, operations, data handling and some capacity.

THE ACTIVITY ON THE HAND GRAPH SHEET
offers children the opportunity to measure their hands.

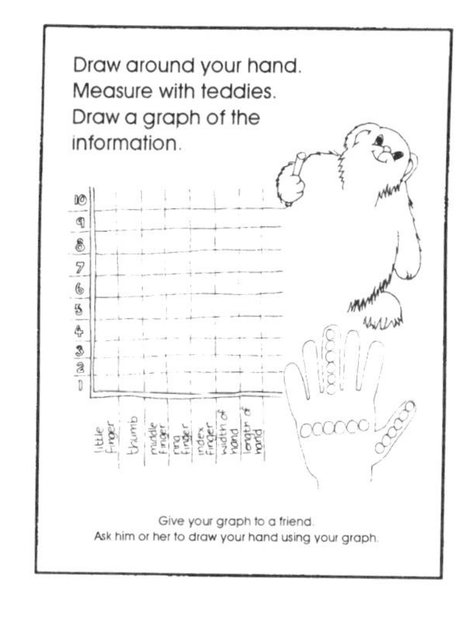

- They make a graph of the measurements and give it to a friend to build a hand.
- From that, a diagram is drawn and compared to the actual hand.
- The graphs and diagrams make a marvellous display and trying to match them up is a fun activity.

FURTHER SUGGESTIONS/EXTENSIONS

- Graph your teacher's hand.
- Make a hand twice the size of yours. Graph it.
- Make a hand half the size of yours;
  quarter the size of yours.
- Make a hand half the size of one of the teachers.
- Make a graph of another teacher's hand. Do not tell the children which teacher you have chosen. Ask the children to find out whose hand it is.
- Make graphs of their feet.

THE ACTIVITY ON THE CYLINDER SHEET
ask the children to roll up a piece of paper to form a variety of cylinders and then fill with teddies

- Children may need to discuss how to roll up the paper to make a thin cylinder.
- Do they roll it straight? Do they roll it so that it grows longer?

FURTHER SUGGESTIONS

- Make a cylinder that will hold the least number of teddies possible and then one that will hold the largest number of teddies possible.
- Change the size of the paper. Change the shape of the paper.
- Which is best, short and fat or tall and thin? Same paper two ways.
- Make a cone.

THE ACTIVITY ON THE TEDDY BAND SHEET
asks children to cover a colour pattern and then continue it.

- Working out the last teddy is difficult because one has to evaluate the space left.
- Evaluate the number of teddies that will fit into that space, then work out the pattern accordingly
- The children will realise how many patterns can be achieved from the same number of teddies.

Put teddies on this pattern.
Continues across your table.
What will the last teddy be?

FURTHER SUGGESTIONS/EXTENSIONS

- This can be done with smaller patterns and less space to start with e.g. a single line of teddies.
- It can be done with more teddies and more space.
- Other types of patterns can be introduced.

  3, 1, 3, 1, 3, 1. three teddies, one teddy, three teddies, one teddy.
  3, 1, 2, 3, 1, 2, 3, 1, 2.

  increasing or decreasing patterns 9, 8, 7, 6, 5, 4, 3, 2, 1, 0 or 9, 7, 5, 3, 1.

  sequences 2, 4, 6, 8, 10 or 1, 3, 5, 7, 9, 11.
- Patterns can be recorded on squared paper.
- Teachers or children may blank out colour and design their own pattern.
- Children can record by colouring the hats on the teddies.

THE ACTIVITY ON THE TEDDY RACE SHEET
gives children experiences with place value.

Blue and green are each 1, red is 10 and yellow 100.

FURTHER SUGGESTIONS

- A simple exchange game where a group of children take it in turns to throw a dice. They collect the number of teddies on the dice. First one to score twenty (2 reds) is the winner.
- Reverse. Start with two reds and give back the number shown on the dice. First one to score 0 is the winner.

THE GAME:

- Teachers can use squares going to 30 or 50 for younger children.
- Teachers may want to do tens and ones only. They can blank out the 100 and value both red and yellow as 10.

Put a teddy on each teddy you can see at the fair.

Guess how many there are.
Now count them.

1. How many teddies:
   On the merry-go-round? [ ]

   On the helter-skelter? [ ]

   Watching? [ ]

2. Which is the largest group? ________ The smallest group? ________

3. What is the difference? [ ]

Draw a fairground picture showing at least 20 teddies.

On each attraction there is a teddy who is the odd one out. Find her. Put a teddy on each.

Put red teddies on the top group of teddies. How many are in the group? Guess then check.

Put blue teddies on the bottom group of teddies.
How many are in this group?

1. Put yellow teddies on the left.
   How many?
2. Put green teddies on the right.
   How many?
3. Where are the least number of teddies?
4. Where are the most teddies?
5. How many altogether?

How is each group different?
Find a way to describe the difference.

Teddies are worth

Red	1
Blue	2
Green	3
Yellow	4

Red	
Blue	
Green	
Yellow	

Put teddies on.
What is the page worth?

1. Change greens to red.
   What is the page worth?
2. Change teddies to make the page worth twice as much.
   Then half as much.
3. Add teddies to make the page worth 50.

Give each teddy a different value.
Make the page worth hundred.

Teacher Notes

The previous sheets involve counting patterns, place value and a variety of experiences with numbers.

THE ACTIVITY ON THE FAIRGROUND SHEET
asks children to place teddies on a variety of attractions at a fairground whilst estimating, checking and dealing with numbers up to twenty.

FURTHER SUGGESTIONS

- Teachers can change the number of teddies to be handled by blanking some out or adding to them.
- Further questions can be posed:

  How many teddies are on the merry-go-round and helter-skelter?
  Make a list of attractions from busiest to least busy.
  Change the teddies around so that none are watching.
  Which is the busiest attraction now?
- If you give a price to each attraction how much has been earned?
- If each teddy had say 20 pence/cents, what could she have a go on?
- What is the largest number of teddies each attraction can hold?

THE TOP/BOTTOM/RIGHT/LEFT SHEET
develops counting skills alongside their work with position.

FURTHER SUGGESTIONS/EXTENSIONS

- If the page is divided in half how many teddies would be on;

  the right side? the left side?
- If the page is divided in quarters how many teddies would be in;

  the top right corner? the bottom left corner?

THE ACTIVITY ON THE SHEET OF TEDDIES CARRYING FLAGS
uses colour as a value.
For example yellow teddy is worth '4', green is worth '3', blue is worth '2' and red is worth '1'.

- Before giving the children the worksheet teachers could do a few simple activities with them.

  What are one red and one green worth?
  What are two blues and three greens worth?
  What is one of every colour worth?
  I have two teddies in my hand worth seven; one is green, what is the other colour?

FURTHER SUGGESTIONS/EXTENSIONS

- Blank out the values given and substitute values to accommodate different abilities or current class projects.

  e.g. 4, 5 and 6. 1, 10 and 100. 2, 4 and 6. (or any tables) 1, $^1/_2$, $^1/_4$ and $^3/_4$.

- Value the teddies.
  Put teddies on the page.
  Calculate value of the page.
  Do not tell people how much each teddy is worth.
  Tell them what the values are and total value of page.
  e.g. teddies values are 3, 4, 7 and 2, total page value is 34.
  Now which teddy is 3, which 4, which 7 and which 2?

- Value the teddies.
  Calculate the value of the page.
  Children have to work out which colour teddies fit on the page.

- Activities can be done with less teddies.

- Use only four teddies.
  Give total value.
  Children work out what each teddy might be worth.

- Use four teddies.
  Give each a value.
  Children work out total value.

- Use four teddies.
  Give value of three teddies and the total.
  Children have to work out 4th value.

On a Sunny Day

Find some paper.
Put teddy in the centre.
Put it in the sun.

Draw around
teddy's shadow

at 10.00am
11.00am
12.00am
1.00pm
2.00pm
3.00pm

Discuss what has happened.

WARNING - NEVER LOOK DIRECTLY AT THE SUN

You need a torch.
Sit teddy somewhere and shine it on teddy.
Shine it on teddy from another place, and another.

What is happening to teddy's shadow?
Can you make it long or short?

Find a window through which the sun is shining. Put teddy in the window.

Use the diagram on the back to record where the sun is at:
10am, 12 noon and 2pm.
Talk with your friends and work out what is happening.

10 am

12 noon

2 pm

Guess how many teddies there are.

Put teddies
on and count.

How many teddies are there?

Guess and count.

Teacher Notes

The previous sheets involve work with the sun and shadows

THE ACTIVITY ON THE WINDOW SHEET
helps children observe how the sun shines from a different part of the sky throughout the day.

- When children are noting where the sun shines from, **they should NOT look directly at the sun, even through sunglasses.**
- By putting teddy in the window your are giving a reference point from which to observe the sun.

THE ACTIVITY ON THE SUNNY DAY AND TORCH SHEET
gives children the experience of looking at shadows and the movement throughout the day.

- Children need to know that light can shine through certain materials and not through others. When light is blocked a shadow is formed.
- After drawing the teddy shadow throughout the day the children will notice that the shadow moves around teddy and that it changes in length.

FURTHER SUGGESTIONS/EXTENSIONS

- Children could make a teddy shadow clock.
- When is the shadow longest?
- When is the shadow shortest?
- When is the shadow as long as your pencil?
- The use of a torch shows the children how shadows change according to the angle of the light being shone.
- Children can keep the torch still (the sun) and move the teddy around.

Teacher Notes

These sheets offer estimation and counting experiences

THE ACTIVITY ON THE PAGE OF TEDDIES
asks children to estimate how many teddies they think are needed to cover the teddies on the page.

- Initially children may guess randomly.They should collect the teddies they think they will need, and record the guess.
- With practice their guesses will get better.

FURTHER SUGGESTIONS

- Teachers can blank out or increase teddies according to the abilities of their children.
- Various questions can be asked:

  Cover five teddies. How many are left?
  Cover three in red and four in blue. How many are covered? Uncovered?
  Cover two more than four. Six less than eight.
  Cover an odd number. Cover an even number. Cover in pairs.

This teddy page contains many more teddies than the previous page, for estimation and counting.

FURTHER SUGGESTIONS/EXTENSIONS

- cover $^1/_2$ of the teddies.
- cover $^1/_4$ of the teddies.
- put teddies into groups of ten.
- put teddies into groups of five.
- put teddies into pairs. How many pairs?
- how many ways can you group the teddies?

Moving teddy.

You might need:
bricks, paper, tape, scissors, string, magnets, paper clips, balloons, glue, elastic bands, plastic bags ....

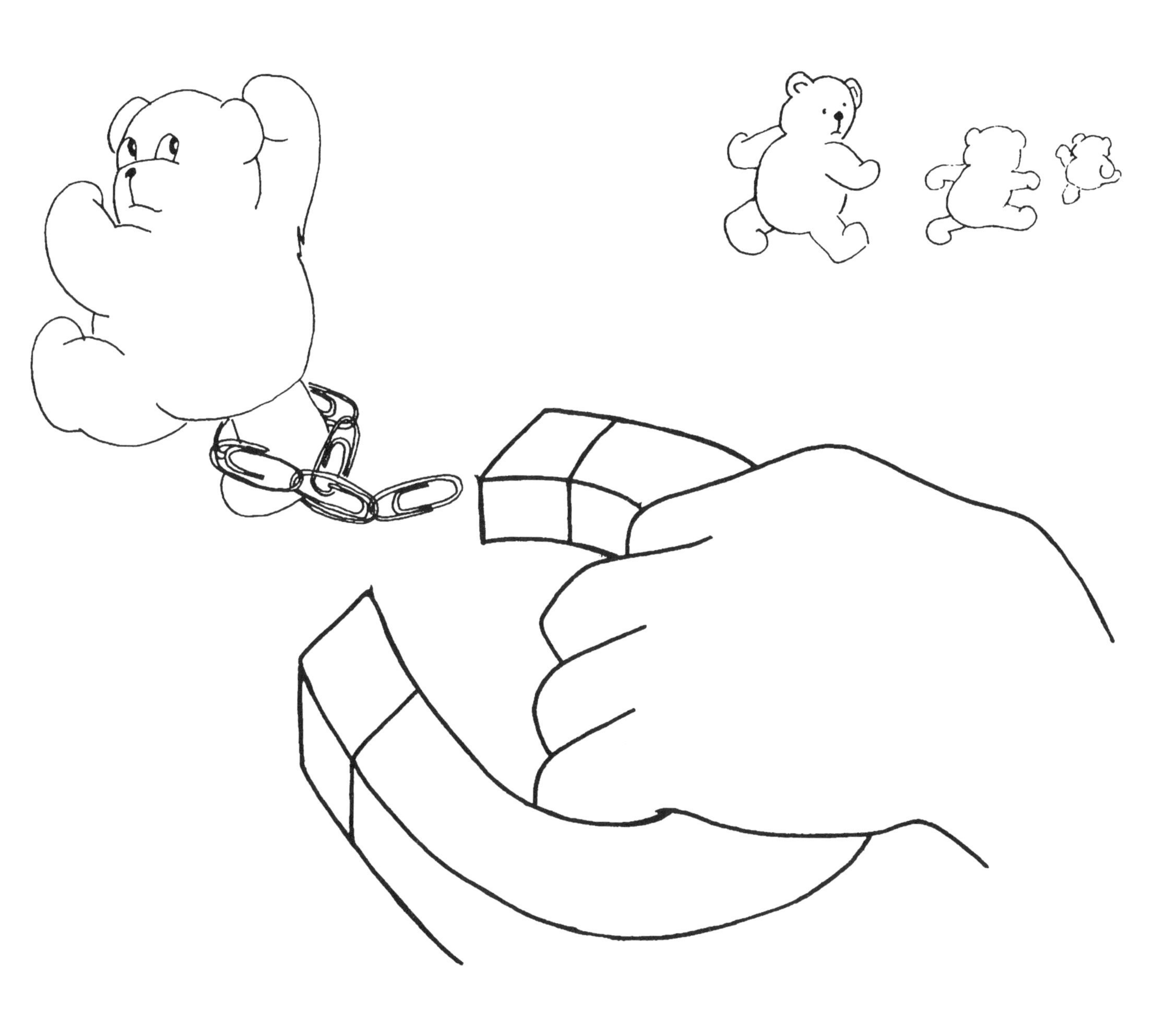

Find ways of moving teddy across the table.
(You can touch her).
Find ways without touching her.
You could use:
air magnets balloons

1. Can you make Teddy float?

2. Can you make your floating Teddy move?
   Use forces such as air or magnets.

3. Can you make your teddy fly?
   With an aeroplane? Without an aeroplane?

All your experiments can be recorded by drawing and writing.

Make a box for four teddies. Value them

red 1 blue 2 yellow 3 green 4

Make the box of teddies worth 10.

1. How many ways can you make a box of teddies worth ten.
2. Find a way of recording this.
3. Make a box of teddies worth fifteen.

I wonder ...
Make up a box worth whatever you want.
Tell your friends how many teddies are in it.
Let them guess what teddies are in the box.

Price your teddies

red 1   blue 2   green 3   yellow 4

Make as many numbers as you can.

You can use + or -   red + green = 4   green - red = 2

Can you fill in the squares with teddies?
The first is done for you.

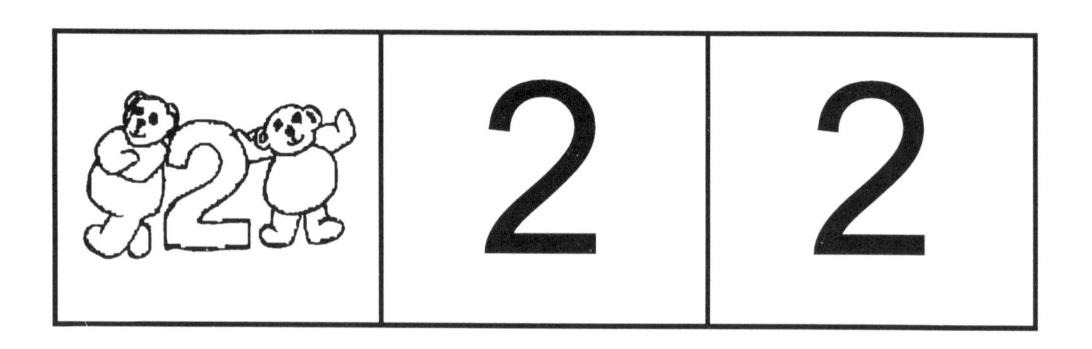

Do the same with the number three in all the squares, with four, with five.
Can you do this with any number?

You can do it with different numbers in each square.
Can you do it with this?

Teacher Notes

The sheets give experiences with forces like air, wind, magnets, pushing and pulling.

THE ACTIVITY ON THE TEDDY MAGNET SHEET
asks children to move teddies using a variety of methods.

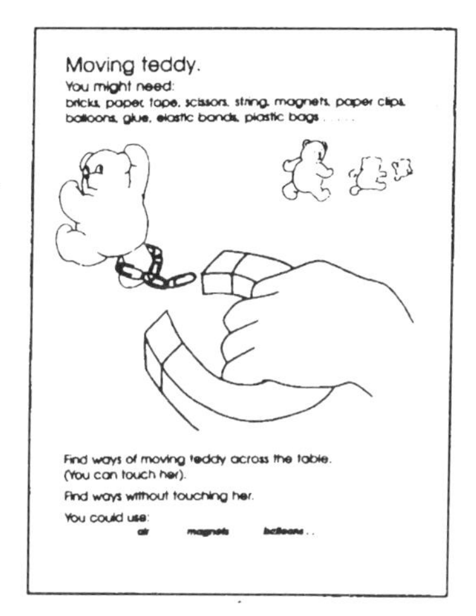

- Initially they will move teddy by pushing or pulling her.
- They can experiment:

  with blowing air.
  air being released from a balloon.
  air from an electric or paper fan.
- How much wind power does one need to move teddy who is

  in a boat; in a plane?

- If a paper clip is attached to the boat or to teddy, a magnet will attract it.
- If a balloon is filled, attached to a boat and released, the boat with teddy in it will move.
- Teddy can be made to float:

  On a raft.
  In a container.
  In a boat.
  By being attached to a filled balloon.
  If the water is so full of salt she cannot sink.

- Can teddy fly:

  By being thrown across the room from a sling?
  By being slid off a surface into air?
  In an aeroplane?
  In a paper plane?
  In a plane and balloon?
  Attached to a balloon?
  In a parachute?
  Being blown by a hairdrier?
  With a bag full of air?

- Make a class book called 'Moving Teddy'.

THE ACTIVITY ON THE SHEET OF TEDDIES CARRYING NUMBERS
allows the child to handle operations as difficult as they can manage.

- Teddies are given values 1, 2, 3, 4. Children are then asked to make up as many numbers as they can using teddy colours as values.

- Numbers can be made up in a variety of ways:

  1 can be yellow (4) and green (3) $4 - 3 = 1$
  6 can be green (3) and blue (2) $3 \times 2 = 6$
  6 can be green (3) and green (3) $3 + 3 = 6$

FURTHER SUGGESTIONS/EXTENSIONS

- The activity can be reversed
  e.g. make up teddy number sentence, red, green and blue. Children work out the sum.

- Use blank squares.
  Put teddies on.
  Work out the value of each square.

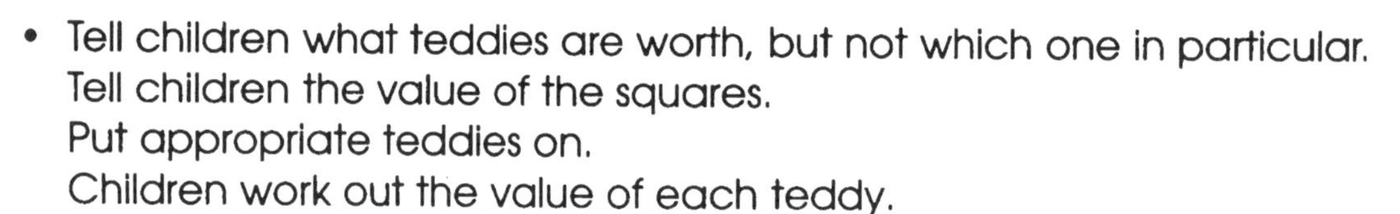

- Tell children what teddies are worth, but not which one in particular.
  Tell children the value of the squares.
  Put appropriate teddies on.
  Children work out the value of each teddy.

THE ACTIVITY ON THE SHEET OF TEDDIES CLIMBING AND FLYING INTO A BOX
encourages children to play with numbers.

Teddies are given a value. Four teddies worth a total of ten are put in a box.

The children we did this activity with made boxes out of interlocking cubes and bricks.

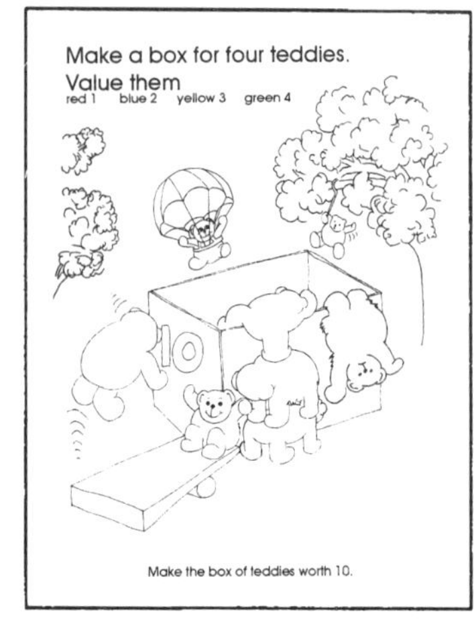

The activity can be reversed.

- The box is worth 12.
  Teddies are worth?
  Who is in the box?

The activity can be done with less teddies or more teddies.

It can be done with letters.

- Give teddy a letter e.g. a, t, e, h, etc . . .
  Put teddies in a box and tell children what the word is.
  What teddies are in the box?

  e.g. hat, ate, tea, eat, hate, the, heat, at.

  Teddies could be sh, th, ch, ph, ea, oa, and so on.

Design a garden for teddies.

Think about gates, fences, a path, trees, seats, a pool.
Anything else?

1. Make your garden. Think about the materials you need, you may need to line boxes with plastic.
2. Fill your box with soil.
3. Plant grass or cress seeds for the lawns. Water every day. Cut when it grows too high.
4. You can make trees by planting fruit seeds like oranges, lemons or acorns.
5. Make a map or plan of the garden.
6. Plan a walk around the garden.

Teacher Notes

This activity involves designing and planning, measuring and constructing a garden for teddies and growing and looking after plants.

- Children make a miniature garden out of an old box or something similar. It is made to suit a teddy so should not be too large.
- The container to make the garden needs to be quite big otherwise the garden itself will be too small. The garden I made with a group of 7 year old children was in a large square metal container, which was about 1 metre square and 15 cms deep. This was a whole class garden. If groups are making one for themselves the boxes need to be smaller. Containers can be as simple as boxes lined with plastic.

  This is an on-going long term project and would need a book or file to be kept for observation maps, sketches and other work associated with it.
- The garden can be as complex or as simple as the teacher wishes.
- Maps/diagrams of the garden can be as complex or simple as the teacher wishes. They could even be to scale.
- Before starting children will need to discuss how to design their miniature garden.
- They might need to visit a few real life gardens.
- Paths, grassed areas, flower beds, fences, walls, lakes, ponds, streams, hedges and orchards are but a few attractions which might be included in their garden.
- Garden seats could be made of clay.
- Fences can be made out of lollypop sticks or twigs.
- Paths can be made out of small stones or gravel.
- Grass areas can be planted with fast growing grass seed, which can be cut as it grows.
- Soil is better and more manageable if purchased from a garden centre rather than taken from the garden.
- Cress seed can be planted in the vegetable patch.
- Beans such as mung will sprout in about three days if kept damp.
- Small cacti and succulents are very slow growing and could be planted in a little cactus soil to make trees. (These should not be watered too much.)

- Seeds from citrus fruit will sprout if planted in plastic bags of damp soil. Seal the bags and store in a dark, warm place for about 4/5 days; longer if necessary. You will see the shoots start to grow. If then planted with a plastic bag over the top, the little shoots will continue growing. After a while, a small plant will form, which can be used as trees. They will need watering and spraying to keep them moist.
  Do not over-water.

- Carrot tops can make lovely sprays and would look good in the garden. They can be trimmed when necessary. If the carrot top is placed in a saucer of water it will start to sprout. Then it can be planted and kept moist.

- Acorns and chestnuts, along with berries will grow but take quite a time to sprout. The acorns and chestnuts will start to grow in the Spring following the Autumn they dropped off the tree. They will grow but need to be left. Children could collect them in the Autumn and plant them in pots outside and leave through the winter. Then plan to use them when the gardens are being designed and made in the Spring/Summer terms.

FURTHER SUGGESTIONS/EXTENSIONS

- A diary could be kept of changes and growth.

- The children can be asked problems like:

  'How long will it take for the grass to grow so that teddy can hide in it '?
  'How many teddies tall will the carrot top grow'?

- Each object could have a symbol and the map have a key.

Make a poster advertising a Great Teddy Event.

Make it exciting.

1. Design and make 'Invitation' cards.

2. Design and make 'Acceptance' cards.

3. Design and make 'Thank you' cards.

Would you like ...

to be a journalist
and interview people
who go to the event.

Start a newspaper
THE TEDDY TIMES.

Make up a story with some friends for teddies to act out.

You need:

a cardboard box,
scissors,
crayons, paint, felt pens,
glue, paper, string.

Make the box into a stage so that teddies can use it.

Think up different scenes.
Make different scenery.

Perform the story with teddies.

1. Make your scenery change.

2. Make teddies move.

3. Design a programme for your play.

After the performance:

Ask the audience to give their opinions about your play.
Ask for an idea on how it could be improved.
Ask for one thing they liked.

Design a 'Treasure Box' for four teddies.

Work out the

measurements,

shape,

materials.

Design a hinge for the lid and a catch to keep it closed.

1. Design a pattern for the outside.

2. Design the inside.

3. Now make your box.

Set up a display of boxes in the class. Now explain to the group how you made your box

Say the things you like about other people's boxes.

Teacher Notes

The previous sheets involve designing buildings, organising and other technology and language work.

THE ACTIVITY ON THE TEDDY POSTER SHEET involves designing and making a poster to advertise an exciting event. A variety of cards are then designed and made, and the event organised.

- Children will need to look at posters and note the eyecatching quality of them.
- Class discussions will need to involve:

  What makes a good poster?
  What is the purpose of a poster?
  What colours stand out?
  How can people's attention be attracted to it?

- To help children design their own cards, a selection of commercially produced cards can be brought from home.
- Class discussions can be held about suitability:

  Why is the colour suitable?
  What is the purpose?
  Is the design good?

- Programmes from other events can be brought into school to be examined.
- Newspapers can be studied for ideas on creating a Teddy Times which could be an ongoing or one off project.
- When a child is the journalist they should plan their questions and use a tape recorder to playback the interview to the class.

  The event might be:
  A play.
  Cake sale.
  Games workshop.
  Teddy bear picnic.
  Story reading (teddy bear stories).
  Any activity from this book.
  Teddy day.
  Everyone brings in soft teddies or similar toys and activities are worked around them.

THE ACTIVITY ON THE TEDDY THEATRE SHEET
is intended for a group of children to work together.

- The box needs to be of a sensible size to accommodate teddies, shoe boxes are quite good and easy to cut.
- Children will need to plan, discuss and talk to the teacher before making their theatre.
- They might need to find books about stages and theatres.
- A simple story should be chosen otherwise the entire project gets too complicated.
- Picture books are good for ideas. The stories are usually simple and the illustrations give ideas for scenery.
- Scenery can be changed by just slipping the next scene over the last one from the top.
- Teddies can be made to move by attaching sticks or straws on to them and pulling or pushing from the side.
- Class discussions are an essential part of this project.

THE ACTIVITY ON THE TEDDY TREASURE BOX SHEET
involves the children designing and making treasure boxes for teddies.

Design a 'Treasure Box' for four teddies.

- Children could bring into school a selection of boxes to look at; chocolate boxes, sweet boxes, jewellry boxes etc,
- Teachers can make nets for them or boxes can be unstuck and flattened.
- Although they may use a ready made box, they will still have to decorate it and make it open.
- The inside is as important as the outside.
- Plenty of material needs to be available. I have found that the children are incredibly ingenious and creative.
- Measuring is important.The box needs to be comfortable for the teddies,not too big nor too small.
- I found with this activity that children wanted to work on their own and produce their own boxes.

FURTHER SUGGESTIONS/EXTENSIONS

- Some children may wish to make other things to go in the box with the teddies. I found that tiny books were very popular.
- Some children wanted to write stories about their boxes, whilst others wanted to write instructions on how to make their boxes.

Teddy Weights.

You need:
foil, string, bags, stickytape.

Put a teddy in a bag. Tie it up.
Use as a 1 teddy weight.

Put five teddies in a bag.
Tie them up.
Use as a five teddy weight.

Put 10 teddies in a bag.
Tie them up.
Use a ten teddy weight.

Find things that weigh:

1 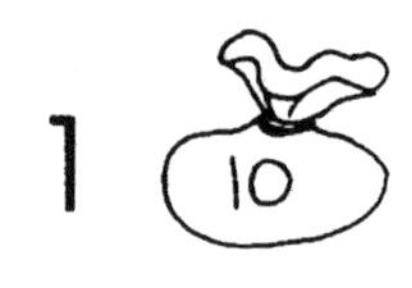 2 

2 6 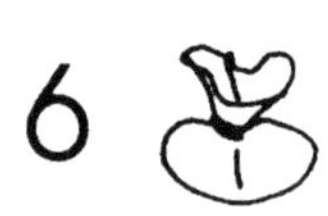

3 4 

Record like this:

The ________ weighs $1_{10}2_{1}$

Make some teddy bread.

Stage 1.
Weigh all these ingredients.

8 teddies of sugar
1/2 teaspoon salt
1/4 teaspoons dry yeast
32 teddies of bread flour
6 teddies of soft butter
1 teaspoon cinnamon
1 large egg beaten well
3 raisins
1 cup milk

Preheat oven to 200c.
Sift the flour into a warm bowl.
Add cinnamon, salt, sugar and yeast.
Make a well in the centre and pour in the egg and milk.
Mix with a warm wooden spoon until the dough comes away from the side of the bowl.
Turn out onto a floured board.
Knead for about 2 minutes.

Stage 2

Punch the dough to get rid of air bubbles.

Divide dough into:

1 ball	5 teddies across (body)
1 ball	3 teddies across (head)
4 balls	2 teddies across (paws)
2 balls	1 teddy across (ears)
1 ball	1 teddy across (nose)

Lightly flatten all these.

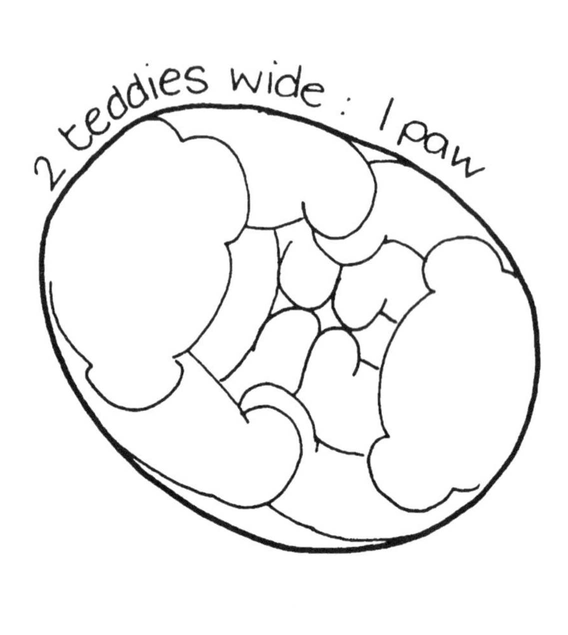

Stage 3

Grease a baking tray.
Put the body in the middle.
Put the head at the top of the body.
Put the nose onto the head.
Tuck ears slightly under the top of the head.
Tuck in the legs around the body.
Snip out fingers and toes with scissors.
Brush the bear with milk and cover.
Leave in a warm place for 40 minutes
until the bear is double the size.

Bake on the middle rack of the cooker for 25 minutes.
Leave and cool and _ _ _ _ _ _ _ _

Teacher Notes

The previous sheets offer work on weight and cooking.

THE ACTIVITY ON THE TEDDY WEIGHT SHEET
gives the children experience with units of measurement and making their own graduated weights out of teddies.

Teddy Weights.

You need:
foil, string, bags, stickytape.

Put a teddy in a bag. Tie it up.
Use as a 1 teddy weight.

Put five teddies in a bag.
Tie them up.
Use as a five teddy weight.

Put 10 teddies in a bag.
Tie them up.
Use a ten teddy weight

- Children will make various units out of teddies

  e.g. a 10 teddy weight; a 5 teddy weight.
- Anything can be used to pack the teddies in: metal foil, plastic bags, old newspaper, cling film.
- The main idea is that children pack say 5 teddies and that becomes 1(5) teddy weight.
- When they make the plasticine weights as heavy as their bags, these are a single weight and the children will understand because they have made them.
- These weights are standard units for them.

THE TEDDY BREAD SHEET
gives the children an opportunity to use teddies as weights when baking bread. They can also use their teddy weights which they have made.

Make some teddy bread.
Stage 1.
Weigh all these ingredients.
8 teddies of sugar.
1/2 teaspoon salt.
1/4 teaspoons dry yeast
32 teddies of bread flour
6 teddies of soft butter
1 teaspoon
cinnamon
1 large egg
beaten well

Flour

Preheat oven to 200c.
Sift the flour into a warm bowl.
Add cinnamon,salt, sugar and yeast.
Make a well in the centre and pour in the egg and milk.
Mix with a warm wooden spoon until the dough comes
away from the side of the bow
Turn out onto a floured board
Knead for about 2 minutes.

- The recipe for dough can take quite a lot of handling, so that each child can knead.
- It is assumed that the activity takes place in small groups, with an adult helping to read the instructions.

FURTHER SUGGESTIONS

- Instead of one large loaf, the dough can be divided between all children. They can then make a smaller teddy bread each.
- If you double the quantities it should be enough for a class of twenty, working in pairs
- The children could work out the recipe for their small bread including the measurements of each body part.
- The cooking process teaches the child that heating changes the dough. It will never return to it's original state such as chocolate or ice.